SCIENCE THROUGH Art

Materials

Andrew Charman

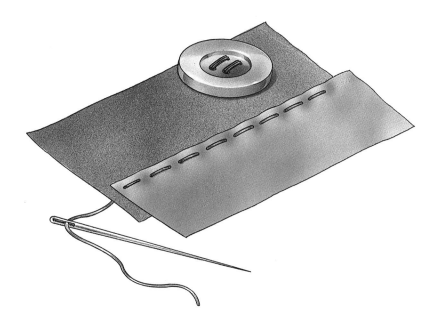

FRANKLIN WATTS

New York/London/Toronto/Sydney

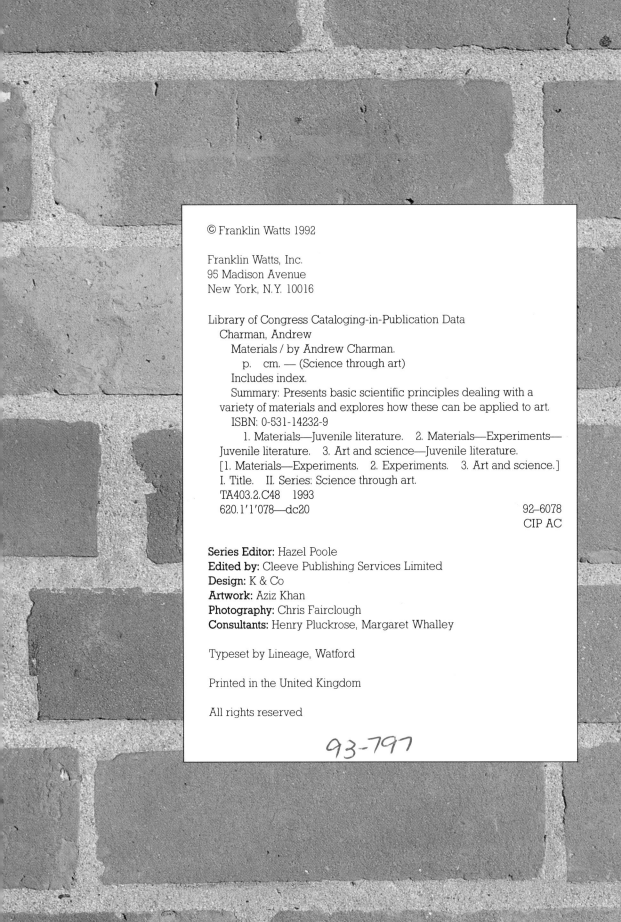

© Franklin Watts 1992

Franklin Watts, Inc.
95 Madison Avenue
New York, N.Y. 10016

Library of Congress Cataloging-in-Publication Data
Charman, Andrew
    Materials / by Andrew Charman.
        p.   cm. — (Science through art)
    Includes index.
    Summary: Presents basic scientific principles dealing with a
variety of materials and explores how these can be applied to art.
    ISBN: 0-531-14232-9
        1. Materials—Juvenile literature.   2. Materials—Experiments—
Juvenile literature.   3. Art and science—Juvenile literature.
[1. Materials—Experiments.   2. Experiments.   3. Art and science.]
I. Title.   II. Series: Science through art.
TA403.2.C48   1993
620.1′1′078—dc20   92–6078
                                                    CIP AC

Series Editor: Hazel Poole
Edited by: Cleeve Publishing Services Limited
Design: K & Co
Artwork: Aziz Khan
Photography: Chris Fairclough
Consultants: Henry Pluckrose, Margaret Whalley

Typeset by Lineage, Watford

Printed in the United Kingdom

93-797

CONTENTS

EQUIPMENT AND MATERIALS

This book describes activities which use the following:

Blotting paper
Bowls (large and small)
Candles
Cardboard (thick, thin, white, and colored)
Compass
Cotton balls
Craft knife
Craft paper
Cutting mat
Dowel rods
Drawing paper
Dyes (for fabrics)
Elastic (thin)
Electric iron
Emulsion paint (acrylic gesso)
Fabric (various sizes and colors)
Felt-tip pens
Food coloring (various colors)
Fun Tak
Hole punch
Inks (various colors)
Jars (or glasses, large and small)

Measuring cup
Modeling clay
Needle
Newspapers
Paintbrushes (various sizes)
Paints
Paper clips
Pen
Pencil
Plastic wrap
Protractor
Push pins
Rubber bands
Rubber gloves
Ruler (plastic and metal)
Spoon
String (fine)
Thread (cotton)
Water
Wax crayons
Wire (florist's)
Wire cutters
Wood (four pieces: 8 inches long, 1¾ inches wide, ½ inch thick)

INTRODUCTION

There are many different kinds of materials in our world. Wood, paper, stone, cotton, wool, plastic, iron, and glass are just a few. Some materials come from nature. Wood, for example, comes from trees, although it is usually treated in some way before we can use it. Plastic is a synthetic material; it is made by man.

Every substance in the world is made up of chemicals. The simplest chemicals are called the elements. They are the simplest because they are made of just one kind of atom. Gold, silver, and oxygen are examples of elements. There are 107 elements in all.

Materials can be liquid or solid, rough or smooth, hard or soft, able to float or sink in water. Some materials can be stretched into different shapes. Others are elastic. They stretch but will return to their original shape. Some can soak up liquids; they are absorbent. Others are water-resistant.

By carrying out the investigations in this book, you will be exploring the nature of materials. Each section begins by introducing you to some scientific ideas. Scientists look at ideas and test them to see if they are true. As you follow the instructions for the activities, you will be testing these ideas just like a scientist.

You will also be an artist. By working through the art activities in this book, you will be using the materials' different characteristics. To make something out of a material you need to know how it will behave. For example, is it hard or soft? Will it float or sink in water? Knowing about materials helps artists to turn their ideas into real things such as paintings and sculptures.

DISCOVERING MATERIALS

Before a scientist starts testing a material, he or she makes a note of what it looks like. What color is the material? Is it shiny or dull? Can light pass through it? Is it light or dark? Is it bulky or flat?

Begin your exploration by gathering as many different kinds of material as you can. You could collect paper, cardboard, wool, fabric, plastic, wood, pebbles, foil, bottle tops, and so on. Some of the materials you will find are natural; others will be "man-made," or synthetic. Most materials can be found in nature, but they are usually changed or treated in some way before we can use them for making things.

Divide your collection into two parts. One set could be of synthetic and "changed" materials such as styrofoam, plastic, molded metal, paper, and fabric. The other pile could be of natural and unchanged materials such as tree bark, pinecones, acorns, feathers, wool, sand, stones, twigs, leaves, and shells.

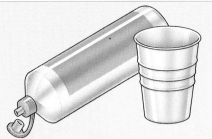

Plastic There are many different kinds of plastic. They are not natural materials but are made by man from oil.

Glass Glass is made by mixing and melting together a number of natural materials in a very hot furnace.

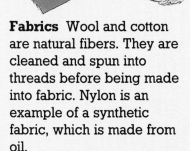

Fabrics Wool and cotton are natural fibers. They are cleaned and spun into threads before being made into fabric. Nylon is an example of a synthetic fabric, which is made from oil.

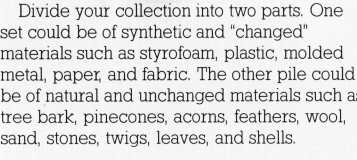

Paper Paper can be made from wood that is turned into pulp and has chemicals added.

Metals Metals are found in nature mixed with minerals in rocks called ore. They have to be heated to high temperatures to be separated from their ores.

Stone Stone is cut out of the earth by quarrying. Sometimes this includes using explosives. There are many different kinds of stone that artists can use in sculptures and other projects.

A material collection

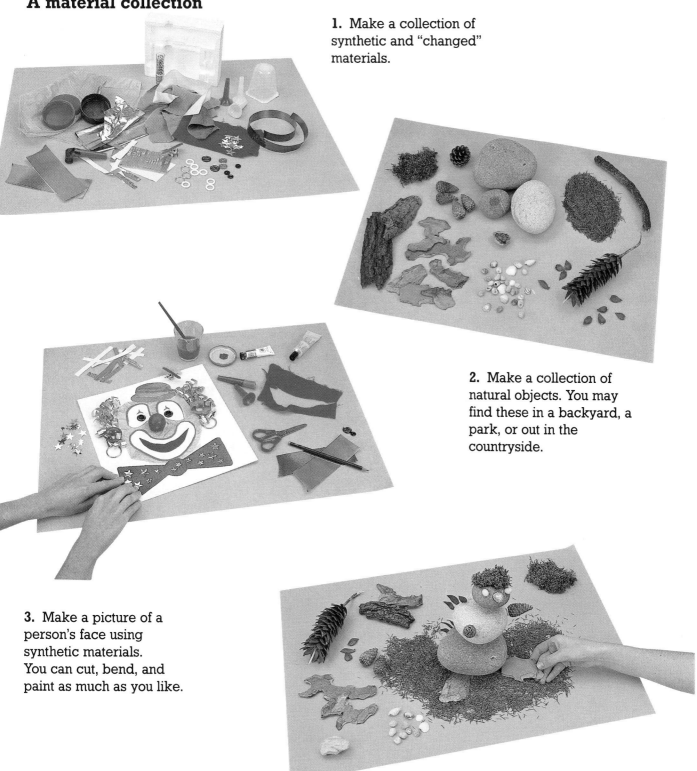

1. Make a collection of synthetic and "changed" materials.

2. Make a collection of natural objects. You may find these in a backyard, a park, or out in the countryside.

3. Make a picture of a person's face using synthetic materials. You can cut, bend, and paint as much as you like.

4. Make a sculpture using natural materials.

ABSORBENCY

Some materials are absorbent. This means that they will soak up, or absorb, another substance. There are a number of different ways of doing this. One of the most common kinds of absorption is when a solid absorbs a liquid. This happens when you spill a drink onto your clothes or when you use a cloth to wipe up something. This is because most fabrics are absorbent.

Blotting paper absorbs liquids. Under a magnifying glass or microscope you can see that there are spaces between the fibers of the paper. Liquids rise up into narrow spaces, even against the downward force of gravity. This is called capillary action. Some solids can absorb gases. Charcoal can absorb ammonia and other dangerous gases. This is why it was once used in gas masks.

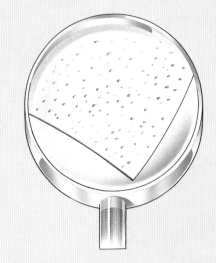

Dark colors absorb more heat than pale ones. This is why people in hot countries often paint their houses white. The white reflects the heat and makes the houses cooler.

Test some of your materials for absorbency by dropping water onto them or by dipping them into water. See which of the materials absorb water. Notice any other features they have in common. What do you notice about the feel of materials that do not absorb water?

Color circle

You will need: blotting paper, cardboard, empty jars or glasses (large ones and small ones), water, a pencil, scissors, paper clips, different colored inks and food colorings, a small paintbrush, glue, and a ruler.

1. Draw and cut out a circle of blotting paper about 8 inches in diameter. Divide this circle into eight segments of equal size and cut them out.

2. Using a straw, put a small amount of two or more colored inks or food colorings in a small jar. Stir them together with the straw. Repeat this eight times with different colors if you want to.

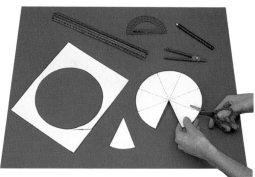

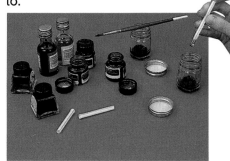

3. Fasten the segments of blotting paper to strips of stiff cardboard with paper clips. Paint a line of one of your mixed inks across the wide end of each segment.

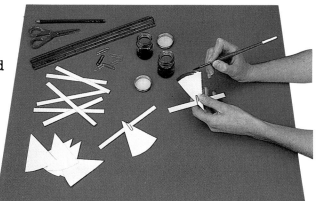

4. Place the strips of cardboard over the glasses or jars. These jars should contain just enough water so that the wide end of each segment is slightly submerged. Now leave the paper to soak up the water.

5. When the colors have reached the sharp end of each segment, remove the segments from the water. Leave them to dry.

6. Glue the segments onto another piece of paper to make the original circle again.

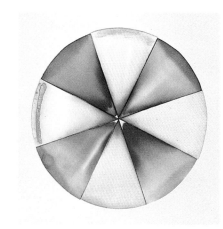

The blotting paper soaked up the water, carrying the inks with it. Some inks are made up of a number of different colors, which travel faster through the paper than others. This is why some of the colors separated. Observe which colors traveled the furthest up the blotting paper. See what happens if you start with a line of just one color. Notice what happens if you remove the segments early. See what happens if you try this experiment with a waterproof ink, such as India ink.

Some materials change shape when you push or pull them. Others bend or fold. This is because they are being stretched. Some materials will stretch or compress and then spring back into their original shape when you stop pushing or pulling. Materials that do this have elasticity. Rubber has this quality. Some metals have elasticity as well, especially when they are coiled into a spring.

Metals are elastic when they are made into wire and then coiled into a spring. However, most materials have an "elastic limit." They will stretch only so far without breaking or staying out of shape for good. They cannot return to their original shape.

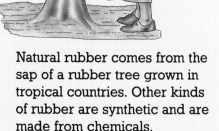

Natural rubber comes from the sap of a rubber tree grown in tropical countries. Other kinds of rubber are synthetic and are made from chemicals.

Bounce a ball, such as a tennis ball, on the ground. It bounces back at you. This is because one side of it is compressed. Its elasticity makes it spring back into its original shape. This makes it bounce back.

Rocket launch

You will need: stiff cardboard, a pencil, scissors, rubber bands, paints, a ruler, glue, and a craft knife.

1. Draw a rectangle, about 6½ inches × 7½ inches onto stiff cardboard and cut it out. Cut a slot running down the center of this shape about three-quarters of the way down. Cut four notches at the top of the cardboard as shown.

2. Draw and cut out a rocket. Then cut out a small rectangle of cardboard and fold it down its center so that it forms a right angle. Make a notch on one of its surfaces and stick the other surface to the back of the rocket.

3. Paint the rocket and its background so that it looks like a space rocket on its launchpad.

4. Join three or four rubber bands together. Attach the ends of this "superband" to the notches at the top of the cardboard rectangle. Slide the rocket into the slot. Attach the middle of the "superband" to the notch on the rocket.

5. Pull down the rocket as far as it will go. The "superband" should be stretched. When you let go of the rocket, the "superband" will return to its original length and launch the rocket into the air.

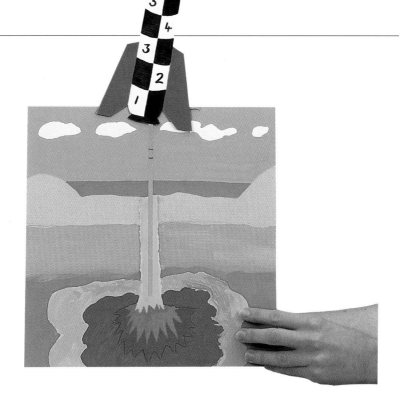

Jumping frog

You will need: a pencil, scissors, stiff cardboard, a length of thin elastic, Fun Tak or similar material, a marker pen, a hole punch, a push pin, paints, and a paintbrush.

1. Draw the outline of a frog onto a piece of cardboard and cut it out. Draw around this shape and cut out another frog.

2. Put the two shapes together so that they match and, with a pencil, mark the outsides with an A. Make a hole through the top of both frog shapes.

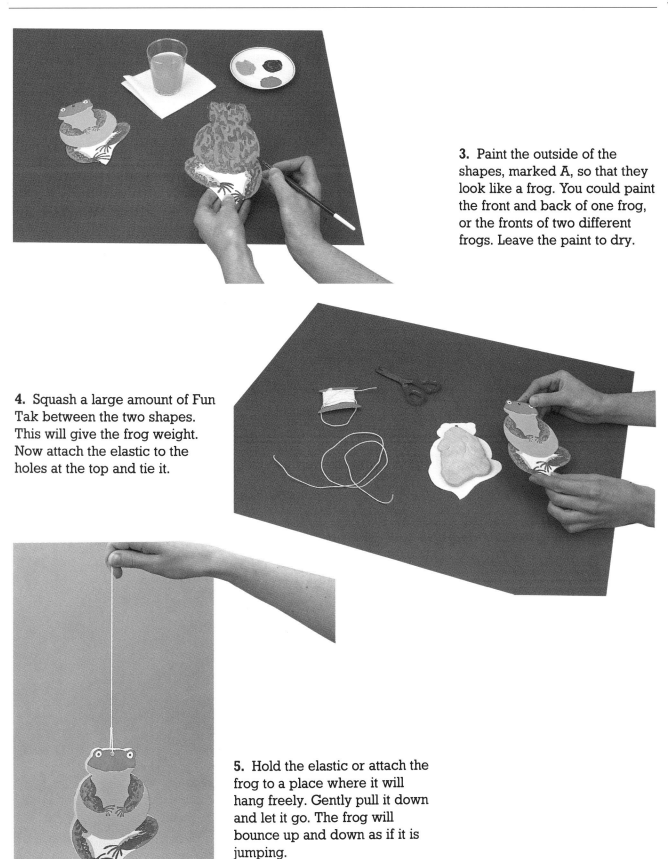

3. Paint the outside of the shapes, marked A, so that they look like a frog. You could paint the front and back of one frog, or the fronts of two different frogs. Leave the paint to dry.

4. Squash a large amount of Fun Tak between the two shapes. This will give the frog weight. Now attach the elastic to the holes at the top and tie it.

5. Hold the elastic or attach the frog to a place where it will hang freely. Gently pull it down and let it go. The frog will bounce up and down as if it is jumping.

HARD AND SOFT

A hard material is one in which the atoms are closely packed and held together by strong bonds. A soft material has atoms that are loosely bonded.

The weight or length of an object can be measured in units such as pounds or inches. The hardness or softness of a material cannot be measured in this way. But you can test the hardness of a material by seeing if it can be scratched or marked by another material.

Diamonds have small atoms which are packed tightly together. They are the hardest known material found in nature. Diamonds cannot easily be scratched. They are used for cutting and grinding other hard materials in many industries.

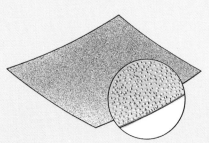

Sand feels soft because the individual grains are so small and fine. It is really very hard. This is why we use sandpaper for smoothing wood and other materials before painting.

The Moh scale is a scale used to test the hardness and softness of a material. The hardest material on the scale is diamond, the softest is talc. Talc is so soft it can be crushed by a fingertip.

Make a collection of materials and test them for hardness. Use a plastic pen cap to try and scratch them. Remember not to scratch anything that someone wants or which may be damaged by your experiment.

Soft sculpture die

You will need: a piece of fabric (16 inches × 12 inches), scissors, a pencil, a bottle top, a needle and cotton thread, glue, cardboard, a ruler, a piece of fabric (6 inches × 6 inches) of another color, and some scraps of fabric or cotton balls.

1. From a piece of cardboard, draw and cut out a square 4 inches × 4 inches. Use this pattern to draw the shape shown in the photograph onto the larger piece of fabric.

2. Cut out the fabric shape. On the smaller piece of fabric draw around a bottle top to make 21 dots. Cut them out and stick them onto the first fabric shape as shown in the photograph. Note that opposite sides of a die add up to seven.

3. Sew three of the edges of the die together. Keep the dots on the inside and leave the last edge unstitched.

4. Before sewing the last edge, turn the die right-side out, so that the dots are now on the outside. Stuff the die with small scraps of fabric or cotton balls. Then sew up the last edge.

5. There you have a soft sculpture die. There are many soft sculptures you can make with fabric. Try making one of a soup can or a banana, or of your hand.

Most of the materials we use for making things are changed from how they were in nature. Some materials are changed before we use them. Paper is made from pulped wood, and cotton is spun into long threads before it is woven into fabric. Other materials are changed as we use them to make things. For example, we mold clay to make sculptures or pottery.

Water and heat are often used to change materials. Water is added to a material so that it softens and can be shaped. Heat is then used to dry out the material so that it stays in its new shape. Sometimes a substance is made by mixing two or more natural substances together. These natural substances are known as raw materials.

Metals are found in nature mixed with other substances in minerals or ores. The metal is separated from its ore by heating it to a high temperature. The molten metal is then poured into a mold to give it shape.

Clay is made up of tiny particles of rock loosely bound together in a mixture that contains quite a lot of water. It is a material that can be easily molded. Once the shape has been made, the clay is fired in a kiln.

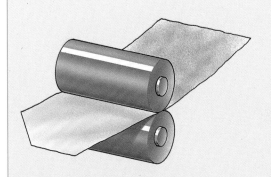

To make paper, logs are first turned into wood pulp. This is made into a runny paste. The paste is spread out in thin layers and dried. The thin layers are rolled into sheets of paper.

The raw materials for making glass are silica, limestone, and soda ash. These are melted together in a hot furnace. The molten glass is blown or molded into the shapes needed.

Papier mâché bowl

You will need: two big bowls, one small bowl, a spoon, old flour, water, newspapers, plastic wrap, emulsion paint (acrylic gesso), colored paints, a paintbrush, varnish, and a measuring cup.

1. Cover the outside of the small bowl with plastic wrap.

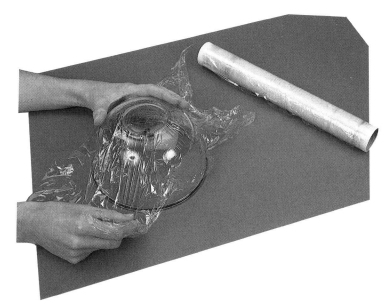

2. Tear the newspapers into thin strips. Tear or cut these thin strips into little pieces and soak them in warm water in the first big bowl.

3. In a measuring cup, make a mixture of flour and water. It should be thick and creamy. Squeeze the water out of the paper pieces. They should be thoroughly soaked and stick together in a lump. In the second big bowl, add the flour and water mix to this lump a little at a time. Squeeze the mixture through your fingers to get rid of any big lumps.

4. Spread the paper pulp onto the outside of the small bowl. Try to make a smooth, even layer. Leave the bowl somewhere warm until the pulp is dry.

5. When the paper pulp is dry, gently ease the pulp bowl away from the glass bowl.

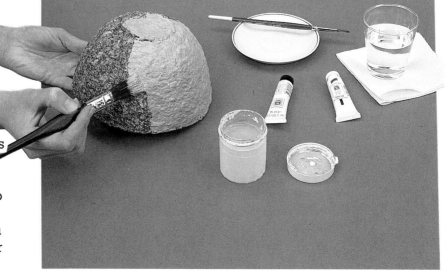

6. Your papier mâché bowl is ready to paint. Begin by covering it in emulsion paint (acrylic gesso) and leave it to dry. Then use your colors. When the colors are dry, you can protect your bowl further by painting it with a coat of varnish.

7. The finished papier mâché bowl.

20 # BENDING AND FOLDING

Many materials can be molded, bent, or folded into different shapes. These materials are very useful to artists. Some materials, like paper, are easy to bend and fold. Other materials, like metals, have to be heated before their shape can be changed. Wire is made of very thin lengths of metal and can be bent easily.

> Make a collection of different materials. See which ones you can change by bending or folding.

Many metals have to be heated before they will bend, but others bend easily. Wire is made of thin lengths of metal especially designed to bend.

Clay can be molded into almost any shape. There are special modeling clays that will harden without having to be fired at high temperatures.

Paper will fold and bend easily. For centuries, artists have used paper to make things. One ancient art of paper folding is called origami.

Pop-up hungry bird

You will need: paper, glue, scissors, a pencil, a ruler, and colored felt-tip pens.

1. Take a rectangular piece of paper. Fold it in half from top to bottom and draw onto it the lines shown here.

2. Cut along the solid line. Fold along the dotted lines backward and forward several times. Open the paper and pull the upper and lower triangles toward you.

3. Mount the cut paper onto another piece the same size, which has also been folded in half. Now decorate the outside and inside of the card. When the completed card is opened and closed, the bird's beak will open and close too.

Tropical fish

You will need: modeling clay or similar modeling material.

1. Modeling clay can be used by an artist in many different projects. You can make a sculpture of a fish, bird, or other animal very easily. Begin by making the main body of the animal.

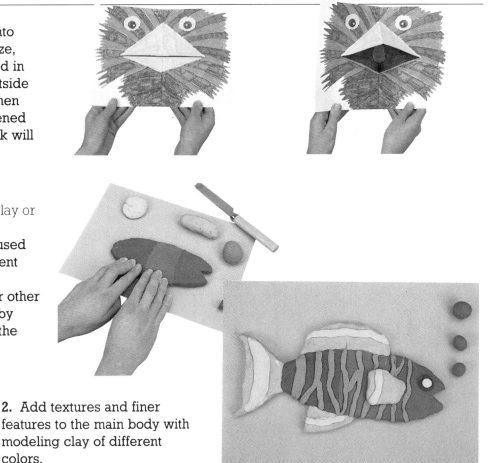

2. Add textures and finer features to the main body with modeling clay of different colors.

Wire sculpture

You will need: florist wire, wire cutters, and modeling clay.

Take a long piece of wire and mold it into the outline of an everyday object. Try to bring the ends of the wire together at the end. Push the ends of the wire into a lump of clay. Try making a sculpture with one length of wire, and then another by joining two or more lengths together.

WARNING: AS SOON AS YOU HAVE CUT YOUR LENGTHS OF WIRE, BEND THE ENDS OVER. OTHERWISE THE SHARP ENDS COULD INJURE YOUR HANDS OR YOUR EYES.

Light comes in waves that we cannot see. Different colors are made up of light with different wavelengths. Red light has the longest wavelength and blue light has the shortest. Sunlight appears to be white, but it is really a mixture of many colors, or wavelengths, of light.

Objects have color because they reflect parts of light back into our eyes. They absorb the other parts. A red object reflects the red part of light and absorbs the others. That's why we see it as red. When we paint a material we are putting a chemical onto it that will absorb some parts of white light and reflect others. Paints come in different colors because they are all chemically different.

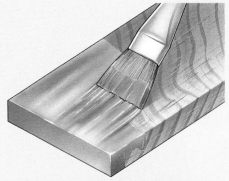

Paints are made up of two things. The color, or pigment, and the liquid that it is dissolved in. This liquid is called the binding medium. It may be water- or oil-based and it lets you spread the paint over a material.

A pigment is the chemical that absorbs or reflects light to give color to an object. The first color pigments that were used by man thousands of years ago were made from earth. Later, people discovered how to use many different colored dyes and pigments from minerals, plants, and animals. Today, hundreds of synthetic dyes and pigments are made.

Fabrics, yarns, wood, paper, and inks can all be colored by dyes. The dye is often mixed with water and then absorbed by the material. Many years ago, dyes were made from plants. Today, there are many synthetic dyes.

Simple batik

You will need: a candle, fabric (white or pale in color), matches, dye, a bowl, rubber gloves, a pencil, craft paper, an iron and scissors.

1. Decide on a design for your batik and draw it lightly onto the fabric with a pencil. Using a candle, drip hot wax onto the fabric where you don't want color to appear. Make sure that the wax forms a thick coating.

WARNING: ASK AN ADULT TO LIGHT THE CANDLE FOR YOU. BE VERY CAREFUL NOT TO BURN YOURSELF OR THE FABRIC.

2. Mix the dye in a bowl or bucket. Follow the instructions on the dye carefully and wear rubber gloves at all times. Dip your piece of fabric into the dye and make sure it is soaked thoroughly. Rinse the fabric and leave it to dry.

3. Remove as much of the wax as you can by hand. Put the dry fabric between two sheets of craft paper. Iron both sides with a hot iron.

When you iron the fabric, the wax will melt and soak into the paper. There should be no color where the wax had been. Most of the fabric absorbs the dye, but the wax makes it waterproof and unable to take the color.

5. You can add another color to your batik. Use the same method described here using a different color.

ROUGH AND SMOOTH

Some of the materials you have discovered have a smooth surface; others are rough, or textured. A textured surface has many uses. One of the most important is that texture increases friction between the object and the surface on which it is standing. This prevents the object from sliding around.

Fingerprints are an example of a textured surface that creates friction and gives grip.

The tires of a car are textured with a tread. This creates extra friction which helps the car to stay on the road.

Make a collection of different materials. Divide the materials into sets. One could be of smooth materials. A second could be of more textured materials, and a third could be made up of really rough materials.

Sculptors create textures to make clay or stone look like fine strands of hair.

Texture collage

You will need: wax crayons, paper, and a pencil.

1. You can make a picture of textures of hard and rough materials, not by actually using them in your artwork but by taking rubbings of them. Take a piece of paper and draw a simple outline of an object or a pattern.

2. Take your piece of paper and the wax crayons and go off in search of textures. Place your paper over any texture you want to use. Rub hard over the paper with the crayons so that the textures come through.

3. Gradually build up your picture by filling in the different areas with different colors and textures. Eventually you will have a complete picture and a record of all the different textures you found.

JOINING MATERIALS

Very few things are made of a single piece of material or just one kind of material. When you make things you will want to join materials together. There are many different ways of joining things, and the ways you choose will depend on the materials.

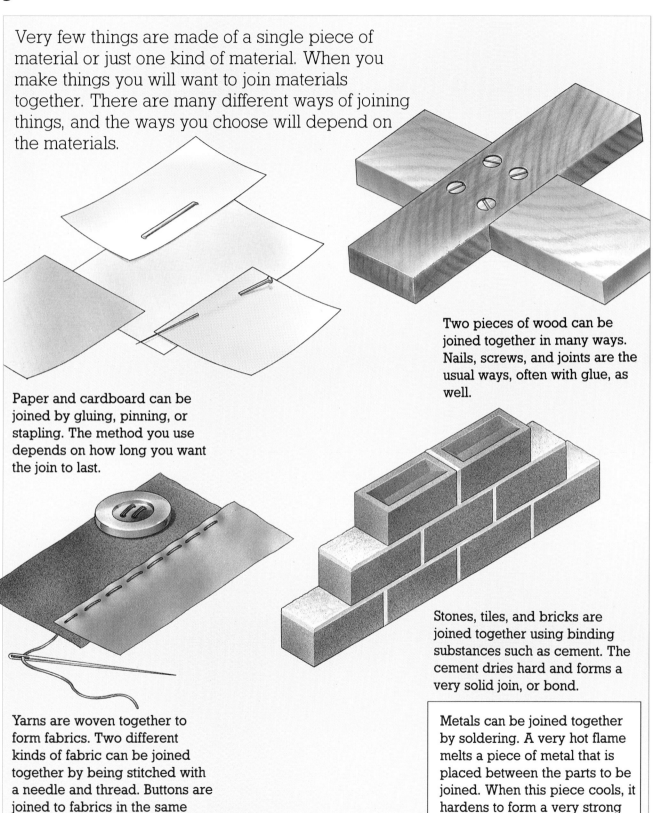

Two pieces of wood can be joined together in many ways. Nails, screws, and joints are the usual ways, often with glue, as well.

Paper and cardboard can be joined by gluing, pinning, or stapling. The method you use depends on how long you want the join to last.

Stones, tiles, and bricks are joined together using binding substances such as cement. The cement dries hard and forms a very solid join, or bond.

Yarns are woven together to form fabrics. Two different kinds of fabric can be joined together by being stitched with a needle and thread. Buttons are joined to fabrics in the same way.

Metals can be joined together by soldering. A very hot flame melts a piece of metal that is placed between the parts to be joined. When this piece cools, it hardens to form a very strong bond.

Make a loom

You will need: stiff cardboard, fine string, strips of fabric, scissors, dowel rods, and a pencil.

1. Draw and cut out a rectangle on a piece of cardboard. This will be your loom. Cut notches along the top and bottom edges. There should be an odd number of notches and they should be opposite each other.

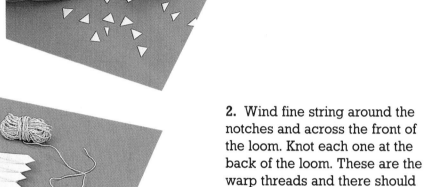

2. Wind fine string around the notches and across the front of the loom. Knot each one at the back of the loom. These are the warp threads and there should be an odd number of them.

3. Weave strips of fabric across the loom, in and out of the warp threads. When you run out of material or want to change color, tie the old strip to the new one and push the loose ends to the back.

4. When you have filled the loom, tie off your last strip of fabric. Lift the weaving off the cardboard. Push a dowel rod through the warp threads at the top and bottom to finish off.

Make a picture frame

You will need: four pieces of wood (8 inches long, 1¾ inches wide and ½ inch thick), cardboard, Fun Tak, scissors, a pencil, a ruler, glue, and a push pin.

1. Take your four pieces of wood and join them together to make a rectangle.

2. On the cardboard, draw and cut eight triangles with sides 2½ inches long.

3. Stick a triangle on each of the four corners on the front of the frame. Put a picture inside the frame, with a cardboard mount in front of it if you like, and place a piece of cardboard behind it. Push a lump of Fun Tak behind the picture in each corner to keep it in place. Stick the remaining four triangles on the corners on the back of the frame.

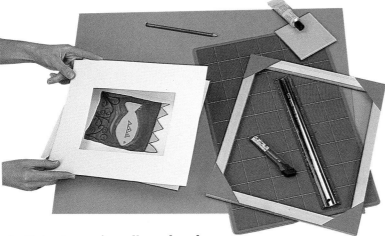

4. Cut a loop of cardboard and attach it to the back of the frame so that the picture can be hung.

5. You could frame the texture picture described on page 26 or any other picture you have made.

Painting fabrics

Squares of fabric can be colored in many ways. You can paint directly on them with water-based paints and a paintbrush. Try painting onto different kinds of fabric and see how they do or do not absorb the paint. You could also try making pictures by sewing together different kinds and colors of fabric.

Man in the moon

The papier mâché method shown on pages 17-19 can be used to make many different things. The man in the moon shown here was made by adding paper pulp to a cardboard shape. Strips of paper can be laid over frames made with wire mesh or wire, or even balloons.

Cardboard sculptures

Pieces of cardboard can be joined together without using glue. They can be slotted together. Take two pieces of cardboard and make slots in each. Turn one so that it is at right angles to the other and slot them together. You can use this basic device to make many different cardboard sculptures, such as the tree shown here.

A photographic study

Make a photographic collection of materials that can be used to make things. You could use some of your photographs to make a collage.

GLOSSARY

Absorption
The soaking up of one substance by another.

Atom
Atoms are tiny particles of matter.

Chemical
Everything is made of chemicals. The simplest chemicals are the elements.

Compress
To press together and force into less space.

Element
A simple chemical that is made of only one kind of atom.

Elasticity
The temporary change in shape of an object when it is compressed, stretched, or bent. When the force is stopped it returns to its original shape.

Fiber
Fibers are the threads that are used to make fabrics and yarns. Natural fibers come from plants and animals. Many synthetic fibers are made from oil.

Firing
Heating to high temperatures in a kiln. Clay hardens when it is fired.

Friction
The force that stops things from sliding past each other.

Molten
Turned into a liquid by heat.

Ore
Rock from which metals are taken.

Plasticity
The change in shape of an object when it is squashed or bent. Unlike an elastic object, a plastic object does not return to its original shape.

Pigments
Substances that are responsible for most of the colors in things. They can be natural or synthetic.

Raw materials
The naturally occurring substances that are mixed together to make something else.

Reflection
When sound and light and other forms of energy are thrown back after hitting the surface of an object.

Sculpture
The art of making figures or designs that have width, height, and depth.

Synthetic
Something which is made artificially by mixing chemicals.

Texture
The surface of a material. You can identify some materials by touch alone.

Wavelength
The distance between two waves. Radio, sound, and light waves come in different wavelengths.